Pocket answer section for
SQA Standard Grade Geography
General and Credit Levels 2004 to 2008

© 2008 Scottish Qualifications Authority Leckie & Leckie, All Rights Reserved
Published by Leckie & Leckie Ltd, 3rd Floor, 4 Queen Street, Edinburgh EH2 1JE
tel: 0131 220 6831, fax: 0131 225 9987, enquiries@leckieandleckie.co.uk, www.leckieandleckie.co.uk

Geography
General Level 2004

1. Answers worth more than 1 mark are indicated by the marks shown in brackets immediately following the statement.

2. A maximum 1 mark will be awarded for appropriate grid reference.

1. (a)

Physical Feature	Grid Square
Deep, narrow valley	2112
Ridge between two valleys, over 500 metres	1808
Broad flood plain	2316
Part of gentle slope, facing south west	1814

(b)

Feature	Letter
Works	B
A467	D
Cairn	C
Scotch Peter's Reservoir	A

(c) Answers might include:
Need to cross River Usk (1); flat, low lying land, so embankments needed (1); Very steep narrow valley, not much room to build road (1); have to make cuttings into the steep side of slope (1); most of the surrounding land is too steep to build on (1)

(d) Answers might include:
If tourist resort, Good place to stay if planning to visit Brecon Beacons National Park (1); people can use town as a base for exploring nearby hills (1); lots of tourist facilities in town eg information centre in 3014 (2); castle museum at 2913 (1); tourists could use leisure centre at 3511 (1).

If market town, Route centre (1); there are lots of farms around Abergavenny (1); such as Mardy farm (2615)(1); which might sell their produce there (1); good roads to transport goods eg A40 (1); largest town for some distance (1); plenty of customers (1).

(e) Answers might include:
Area A: coniferous forests, which some people think might spoil the scenery (1); and stop farmers grazing their sheep (1); electricity pylons spoil view (1); may be access restrictions around reservoirs (1)

Area B: spoil-heap doesn't look good (1); may also be a danger for people walking (1); could pollute the ground (1); cause hazard for wildlife (1)

1. (f) Answers might include:
Advantages: edge of built up areas, land likely to be cheap (1); on main A class road (1); so goods can be transported easily/quickly (1); may be plenty of workers available since old industry has declined (1)

Disadvantages: difficult to expand (1); sloping land to north (1); marsh to north (1); wooded areas to east and south (1); which may need to be cut down (1)

2. Answers might include:
Hard rock more resistant (1); so not so quickly eroded as soft rock (1); plunge pool eroded by fast flowing water (1); which undercuts overlying hard rock (1); which eventually collapses (1); filling up plunge pool with debris (1); this ongoing cycle causes waterfall to move upstream (1); process is repeated over years (1).

3. (a)

(b)

Location	Weather System
British Isles	Depression
Spain	Anticyclone

British Isles has fronts (1); isobar values over Britain lower than 1000mb (1); getting rain (1); also windy (1); whereas isobars over Spain over 1000mb (1); and far apart (1); Spain also has obscured sky, fog being common in anticyclones (1).

4. (a) Answers might include:
Temperature stays high all year, varying from 25 to 30°C (2); hottest month is May (1); temperature range is 5°C (1); Rainfall is heavy (1); with two main peaks, February and September (1).

Geography
General Level 2004 (continued)

4. (b) Answers might include:
Hillsides lack vegetation cover, so will be prone to soil erosion and landslides (1); heavy rain-fall will cause gullies to develop in steep slopes, accelerating soil erosion (1); habitats are destroyed, causing animals to lose their homes (1); and become rare/extinct (1); clean rivers become choked with silt from soil erosion (1); Choked rivers flood (1); build-up of CO_2 in atmosphere (1); reference to possible effects of changes on indigenous people (1)

5. Answers might include:
Rough grazing: found in highland areas (1); where rainfall is highest (1); and temperatures are low (1); these areas are too difficult for arable farming due to steep slopes (1); but are suited to sheep (1)

Barley: in lowlands (1); where temperatures are warmest (1); and rainfall the least (1); in lowlands, land is flatter and more suited to mechanisation (2); crops grow better in warm conditions (1)

6. (a) Answers might include:
Techniques: Pupils could go to the local library and find information from old maps (1); they could look at old newspapers for photographs of the area in the past (1); they could interview older residents and ask questions about how the area has changed (1); do a land use survey of present day landscape (1); plot current land use on land use map (1); take photographs of area (1)

Reasons: old maps could be compared with new ones to see changes (1); old photos would help them to see what the area used to be like and they could compare these with their own fieldwork to see how Inverfirth has changed (2); older residents would give them information about the changes in the area (1); Land use maps provide up-to-date dates for comparison with 1974 (1); or any other valid answer.

(b) If **Yes**, answers might include: More leisure facilities now (1); eg marina and picnic sites (1); less pollution from factories (1); and from untreated sewage (1); area looks better now as spoil heaps have been landscaped and the beach cleaned up (2); smoky power station replaced by mining museum (1)

If **No**, answers might include: factories have closed down, so many jobs will have been lost (1); families have to move away (1); no caravans for people who may have wanted to stay next to the beach (1); Although there is a new motorway, there may be a lot of congestion on it as railway line has been taken up and so goods have to go by road (2).

7. (a) The following is a model answer. Site A or site C could be chosen, so long as valid reasons for the choice are given.

If **B** chosen: Close to motorways and main roads (1); for easy access for deliveries and customers (1); away from congested centre of town (1); outside town where land is cheaper (1); lots of room for building/expansion (1); room for car park (1); Greenfield site with pleasant environment (1); away from existing CBD and shopping centres (1); Site A not as good because it's close to other centre (1); congested, makes transport difficult (1); or in expensive land near CBD (1); Site C not as good because.....

(b) **Techniques**: line graphs (1); multiple line graph (1); bar graph or pie charts (1); series of bar charts or pie charts, one for each year (2)

Reasons:
Line graph shows trends over time, from 1980 to 2000 (1); can show rate of changes by steepness of line (1); can use different colours of lines for each employment category to distinguish them (1)

Pie charts: employment statistics can be converted to percentage for each year (1); size of sectors proportional to number of workers (1); can be enhanced by colour (1)

Bar graph: height of bar proportional to number of workers in each sector (1); making comparison easier (1); can be enhanced by colour (1).

8. (a) Answers might include:
Compared to the overall population, Nigeria has a much wider base than the USA (1); Nigeria's population drops rapidly after the age of about 30 whereas the USA's begins to increase (1); forming a 'bulge' in the 40s age range (1); a much bigger group of people survive to old age (ie over 65) in the USA than in Nigeria (1).

(b) (i) **Nigeria**: Too many young people, so may be shortage of teachers (1); they will grow up and have more children (1); population will continue to grow fast, putting strain on resources (1).

(ii) **USA**: People currently in middle age will retire and cause drain on resources (1); USA will have shortage of labour if not enough young people coming through (1); costs for medical care will increase (1); increasing number of old people and not enough people to look after them/fund their care (1).

9. EU farmers get lots of advantages compared to farmers from Mozambique (1); eg money from their government to grow sugar beet, unlike farmers from Mozambique who get no support (1); because EU buys sugar in bulk they probably only pay Mozambique a small amount for their sugar (1)/Mozambique has to sell sugar at low prices (1); Mozambique has to pay an extra cost (tariff) to export sugar to the EU; they cannot compete on equal terms (2); they cannot export their refined sugar which would earn them more money than raw sugar (1); EU farms will be more efficient since they are mechanised, unlike farms in Mozambique that rely on labour (2); or any other valid point.

Geography
Credit Level 2004

1. (a) The following is a model answer. Several squares could be chosen, so long as valid reasons for the choice are given.

    eg if **9771** is chosen, answers might include: Main roads converge here (1); congested unplanned layout (1); many churches and a cathedral (1); police headquarters (1); court (1); old buildings such as the Bishop's Palace and castle (2); University (1); bus station (1); tourist information offices (1); markets (1).

    eg if **9770** chosen: answers might include: Railway station (1).

   (b) Answers might include:
    B is closer to the CBD so is more congested (1); B is laid out in a grid-iron pattern whereas A has a more varied pattern of geometric curves, crescents and cul-de-sacs (1); this road layout makes A safer than B (1); there is no industry near A but there is evidence of large industrial buildings beside B (1); A appears to have more recreational and ornamental open space in and near it than B (1); creating a more pleasant environment (1); all this suggests that A is a newer residential area than B (1).

   (c) Answers might include:
    Questionnaires could be issued to residents of Bracebridge Heath (1); If a large enough number of responses is obtained it should provide a representative sample (1); If questionnaires are kept brief and simple to use (eg tick boxes) then a high level of responses should be obtained (1); and the information will be easy to process (1); Description of technique (1).

    Interviews could be conducted with people living in Bracebridge Heath (1); More detailed information can be gathered in interviews than by using questionnaires (1); Local people can be expected to know where large numbers of other locals (their friends or neighbours) travel to work (1).

    A traffic count could be carried out on the main road between Bracebridge Heath and Lincoln (1); Doing this in the morning and evening would reveal whether or not many people leave Bracebridge Heath in the morning and return in the evening (1); Only one main road is involved so this could be easily organised (1).

   (d) Answers might include:
    9771 is the oldest part of the town so contains places of historical and cultural interest such as the castle and the cathedral (2); and services for tourists such as the bus station and information centres (2); This area is built-up so has little scope for tourism based on outdoor activities (1).

    9468 and **9469** are nearer the edge of town with scenic lakes and woodland (1); which is suitable for the creation of a country park (1); with a variety of recreational opportunities including

Geography
Credit Level 2004 (continued)

1. (d) (continued)

sightseeing, picnics and birdwatching (2); Land is less expensive so there is room for camping and caravanning (2).

(e) Answers might include:
Land is suitable for the use of machinery because it is gently sloping (2); and the fields appear to be large (1); Gentle slope means land is well drained (1); Farms in the area are well spaced out so appear to be large which is typical of modern arable farms (2); as are the regular field shapes (1); Close to a big market in Lincoln (1); or any other valid point.

(f) Answers might include:
Much of the flat land beside the river has not been built on (1); eg at 966686 (1); because of flood danger (1); Most of the remaining land on the valley floor has roads and housing built (1); because it is flat land suitable for building (1); The valley runs N/S so the roads run this way as well (1); The river is a barrier to E/W communications with only one bridge in 3 kilometres (2); Flat land not suitable for building is commonly used for allotments (1); Building is restricted in the E of the area by steep slopes (1); The sloping land is sometimes used for recreation instead eg golf course (2); The well drained plateau/flat land above the slope is used for settlement or farming (1); or any other valid point.

2. (a) Answers might include:
Pyramidal peak: During the ice age snow collected in hollows around a mountain (1); As the snow piled up the lowest layers turned to ice (1); This eroded the mountain on all sides (1); creating deep corries (1); These got progressively bigger and their back walls were eroded back towards each other (1); producing a steep-sided peak between them (1).

Credit will be given for mention/explanation of plucking, abrasion and frost shattering.

(b) The following is a model answer.
for **HEP**, answers might include:
Deep, narrow valleys (1); ideal for water storage (1); Abundant rainfall/snowmelt (1); Dam site in hanging valley provides head of water above power station (2); Hard rock to support dams (1); Impermeable rock which doesn't allow water to seep (1); or any other valid point.

3. Answers might include:
The heavy rain associated with the warm front dies out as the front moves NE (1); While in the warm sector there will be mild, partly cloudy weather with occasional drizzly showers (2); When the cold front arrives there will be heavy rain and possible thunder (2); Once the cold front has passed the temperature will fall, skies will clear

3. (continued)

and rain will stop (2); Due to the changed orientation of the isobars wind direction will change from SW to W (2); then NW (1); or any other valid point.

4. Answers could include:
In order to reduce pollution the UK government/Scottish Parliament could introduce new environmental laws to prohibit the discharge and dumping of untreated sewage (1); The EU already has legislation to reduce this (1); To stop overfishing the EU has imposed strict quotas on whitefish catches (1); and fishing boats are only allowed to fish on a fixed number of days each month (1); Fishery patrol boats are used to enforce the regulations and heavy fines are given to the fishermen who break them (2); Organic fish farms help to reduce the threat to surrounding sea life (1); Marine conservation areas could be set up to protect whales etc from the effects of seismic testing and oil exploration (2); or any other valid point.

5. (a) Answers could include:
A has more livestock than B because the land is higher (1); and so will be colder and wetter than B and thus less suited to crop growing (2); Since A has a much bigger range of altitude than B slopes are likely to be steeper and thus less conducive to the use of machinery (2); On B there are a large number of part-time workers needed at harvest time (1); B is much smaller than A, so, to make a living, the land must be farmed more intensively with a higher yield per hectare (2); and by planting high value crops rather than grazing sheep which require large grazing areas (2).

(b) **Techniques**: Pie graph (1); divided bar graph (1); bar graph (1).

Reasons: Both pie graphs and divided bar graphs show proportions and the figures are in percentages (1); Since the size of the slice (or bar) is proportional to the percentage the major land uses can be quickly identified (2); Each can be enhanced by the use of colour (1); to allow easier comparisons to be made (1).

6. Answers could include the following:
Greenfield site is cheap land (1); Flat land for building (1); Easy access to main roads (1); for movement of products and parts (1); and for access by workers (1); Skilled labour supply available due to closure of British Railways Engineering (1); Component parts like tyres made nearby (1); approx. 45 km away (1); cutting transport costs (1); Nearby large cities like Derby and Nottingham provide labour force and market (2); International airport is only 60 km away allowing Toyota senior management easy access (2); In open countryside so room for expansion (1).

7. (*a*) Answers might include: Main pattern is from peripheral south and east of Europe to the core north of W Europe (2); Concentration in Germany and the low countries (1); Migrants from each country of origin choose to settle in different parts of W Europe (1); Immigrants from the Caribbean, Pakistan and India arrive in Britain (1); N African immigrants choose France or other areas of continental Europe (1); a third group from Turkey and the Balkans, together with Iraqis arrive in mainland countries eg Germany (1); Within EU, people move from Eire to UK (1); or from Portugal and Spain to countries further north (1).

(*b*) Answers might include:

Advantages: migrants have chance to get better jobs (1); earn money for their families (1); feel safer/more secure (1); escaping from war/persecution (1); better chance of schooling (1); have access to better health and social services (1).

Disadvantages: may be detained in asylum centres (1); children may not get proper education in local schools (1); local people may resent them (1); risk of violence/discrimination (1); have difficulty getting housing (1); may be moved to housing a long way from other migrants (1); short-term permits so may have to leave after a few years (1); language difficulties (1).

8. Answers could include:

The new member countries will have easy access to a huge market within the EU, which will give their industries the prospect of increased sales and profits (2); They will not have to pay tariffs on goods to the rest of the EU (1); Their citizens will be able to look for work in other EU countries and so earn higher wages which they can send home (2); The new member countries will benefit from increased bargaining power in negotiations with other trading blocs (1); They will be able to join the Euro so people will not have to change currency when travelling to the other Euro states (2); Each country will have the benefit of 24 other allies if there is a threat of war (1).

9. Answers could include:

Aid organisations such as the Red Cross would have flown in medicines to help check the spread of disease (1); because lack of clean water might lead to outbreaks of cholera etc (1); Tents, blankets and temporary accommodation would also be needed because of all the homes that were destroyed (1); and refugee camps would need to be built (1); Emergency food supplies would be sent as shops and markets in the villages would have been affected by the lava flows, leaving people with nowhere to buy food (2); Also, because farmland and crops were destroyed, people in the countryside would need these supplies to live off (1); As Congo is a poor country organisations such as the UN would send money to help rebuild roads, airports and other facilities destroyed by the lava (1).

Geography
General Level 2005

1. (*a*) (i) *3 marks for all 4 correct, 2 marks for 2 or 3 correct, 1 mark for 1 correct.*

Glacial Feature	Grid Reference
U shaped valley with misfit stream	7796
Corrie with lochan	6301
Hanging valley	6399
Truncated spur with crags	6699

(ii) *1 mark per valid statement, 2 for a developed point. 3 marks may be awarded for a good labelled diagram(s).*
e.g. U-shaped valley with misfit stream: glacier erodes a normal V-shaped valley (1); sides are made steeper and valley bottom is deepened by abrasion and ice plucking (2); once the ice has melted the misfit stream which is left looks too small for the large U-shaped valley (1).

(*b*) *Up to 2 marks for valid uses for the flood plain; up to 2 marks for valid ways in which the problem has been overcome.*

(i) Answers might include:
- the flood plain has been used for a golf course (1);
- farmland (1);
- as a transport route for the roads (1);
- and railway line (1);
- as a nature reserve (1);
- housing (1);

(ii) Answers might include:
- in places the roads and railway have been built up on embankments (1);
- artificial banks have been built up on either side of the Spey (1);
- most of the buildings are just above the level of the flood plain to avoid flooding (1).

(*c*) *No marks for choice.*
Answers might include:
- LIVESTOCK: steep land so crops cannot be grown (1) or machinery used (1) as farm is on steep land soil will be poor (1) so rough grazing will be found and sheep can survive on this (1).
- MIXED: all of the above but also fodder crops may be grown close to farm on gentler slopes (1).
- ARABLE: on floodplain (flat land) (1).
- Accept any other relevant point.

(*d*) *3 marks for all 4 correct, 2 marks for 2 or 3 correct, 1 mark for 1 correct. Accept correct grid references.*

Geography
General Level 2005 (continued)

1. (*d*) (continued)

Description of Site	Name of Settlement	
Between two tributaries to the north of the River Spey	Newtonmore	7199
On land sloping gently to the north-west surrounded by forest	Insh	8101
On the floor of a U-shaped valley next to a tributary of the River Calder	Glenballoch	6799
A tributary of the River Spey runs through the middle of this settlement	Kingussie	7500

(*e*) *For full marks, must use map evidence. No mark for choice. Maximum (1) mark for grid references. Accept list.*
> Possible answers may include:
> **YES**
> * Footpaths for walkers at 6999 (1)
> * camping/caravans at 712982 (1)
> * picnic site (1)
> * tourist information centre (1)
> * parking (1)
> * viewpoint (1)
> * toilets (1)
> * walks/trails (1)
> * Aqua Theatre (1)
> * Folk Museum (1)
> * Hotel (1)
> * golf course at 7198 (1)
> * Places of varied scenery and footpaths from parking places at 692998 (1)
> * Water sports on River Spey.
> **NO**
> * Lack of accommodation (1)
> * unreliable weather (1)
> * marked footpaths steep (1)
> * few areas accessible by car (1).

(*f*) *Mark 3:1, 2:2 or 1:3. 1 mark per valid point, 2 for a developed statement.*
List of landscape features max 1 mark.
> **Advantages:**
> * it is on marshy land which can't be used for much else (1);
> * there is a variety of habitats to suit different animals (1);
> * there are roads nearby so that people will be able to get there easily to see the wildlife (1);
> * close to mountains and forest so there will be a lot of wildlife in the area (1).

1. (*f*) (continued)
> **Disadvantages:**
> * several roads run around the nature reserve so traffic may be a danger to wildlife and exhaust fumes may cause pollution on the reserve (2);
> * it might be too easy for people to get to, so they might disturb creatures on the reserve (1).
> * Sewage from nearby towns such as Kingussie will have to be carefully treated so that the nature reserve is not spoiled (1).

2. *1 mark for each point that explains the formation. 2 marks for a developed point. A suitably labelled diagram could be awarded full marks.*
> * Softer or less resistant rock is worn away/eroded (1).
> * Harder rock is undercut (1)
> * the unsupported rock collapses (1).
> * A plunge pool is formed at the bottom of the waterfall by the force of the waterfall (1).
> * Accept any other valid points.

3. (*a*) *No marks for choice. 1 mark per valid reason, 2 marks per expanded point. Accept negative points about the other locations.*
> Possible answers may include:
> **YES**
> * Quite near the Geography Department for easy access (1)
> * out in an open space (1)
> * should get direct sunlight from the south (1)
> * allows sunshine recorder to work properly (1)
> * B is sheltered by the trees so instruments would give false readings (1)
> * rain could drip into the rain gauge (1)
> * C is too far away from the Geography Department (1).
> **NO**
> * On tarmac so rain gauge could not be dug into the ground (1)
> * pupils playing in the playground might interfere with the weather instruments (1)
> * football might knock instruments over (1)
> * quite sheltered by buildings on two sides (1)
> * C is better because it is on grass (1)
> * so easier to locate/fix instruments (1)
> * car fumes heat might affect temperature readings (1).

(*b*) (i) *1 mark per valid description*
> Possible answers might include warm (1) dry (1) sunny (1).
(ii) *1 mark per valid explanation.*
> Possible answers might include:
> * Cold because coming from polar areas (1)
> * wet because travelling over oceans (1)
> * Polar maritime air mass (1).

4. *(a)* 1 mark for each climatic area correctly identified.

Area	Name of Climate Region
A	Tundra/Cold desert
B	Mediterranean
C	Equatorial/Tropical rainforest

(b) 1 mark for each valid point. 2 marks for a developed point.
- Failure of rainfall (1) high evaporation rates (1)
- Cattle and sheep/goats/camels overgrazing pasture (1)
- Cultivation on land when rainfall is low (1)
- Clearing of trees, shrubs for firewood (1)
- Population pressure on marginal land (1).

5. *(a)* No marks for choice of land use.
For any chosen land use answers could include as appropriate:
- Land is cheaper away from city centre (1);
- plenty of room to build/expand (1);
- good access via motorway/ring road (1);
- pleasant environment away from congestion/pollution in city (1);
- space for car parking (in leisure or retail areas) (1);
- on cheaper land houses can have gardens (1).
- List – no more than 3 – 2 marks.
- Accept any other relevant points.

(b) 1 mark for each relevant technique. 2 marks for valid reasons.
Answers could include:
Technique:
Issue questionnaires to a large number of customers (1).
Reason:
If sample interviewed is large enough, information is likely to be representative of whole population (1).
Information gained of different types, e.g. age groups and reasons for use (1).
Technique:
Interview shop owner/manager (1).
Reason:
He/she is likely to have good knowledge of area served by store (1); from customer addresses/deliveries (1); if questions are kept brief and supplied in advance co-operation is more likely (1).
Technique:
Count the cars arriving in the car park (1)
Reason:
If done at different times/on different days will allow pattern of use to be identified (1); using a worksheet with space for tally marks will help keep count accurate (1); results can be compared with results from other areas (1); to judge how successful centre is (1).

6. No marks for choice. 1 mark for simple point, 2 marks for a developed point. Accept yes/no answers
YES
- Organic crops are healthy (1) because they are not contaminated by chemical fertilisers and pesticides (1).
- Chemical fertilisers enable farmers to produce more crops (1) and increase their profits (1).
- GM crops can produce better yields (1) and be developed to be resistant to diseases (1), decreasing the need for pesticides (1), thus reducing pollution of streams/rivers.
- Larger fields enable machinery to be used (1) increasing efficiency (1) and saving money by not having to pay workers (1).
- Diversification means that farmers are less dependent on just producing food for their money (1) and are consequently more secure financially (1).
NO
- Organic crops are more expensive to produce (1) and consequently cost more to buy (1) so that they find it difficult to compete with other conventionally produced crops (1).
- GM crops have not yet been proved safe (1). There is a risk that the crops could be contaminated (1) by pollen being blown (1) and cross-fertilisation (1).
- Larger fields have hedgerows to be cut down (1) damaging wildlife (1).
- Any other valid point.

7. No marks for choice. 1 mark for simple point. 2 marks for a developed point. Accept negative points about the alternative.
A
- Has a skilled workforce available since steel produced here only 5 years before (1) and the ex-steel workers may not have been able to find alternative work (1).
- There is a transport infrastructure in place (1) with a main road nearby and a rail link to the coast to enable iron ore to be transported easily to the steelworks (1).
- There is still coal available in the deeper seams (1).
- New roads/railway need to be built at B (1).
- There will be derelict land available for building (1).
- Reclaiming marshland at B might be opposed by environmentalists (1).
B
- The iron ore terminal can be re-opened (1).
- Deepwater enables coal and iron ore to be imported (1).
- Transport costs at A will be greater (1) since the local iron ore is exhausted and has to be imported (1).
- Marshland is not presently being put to any economic use (1) so it will be cheap to buy (1).
- Land is flat and easy to build on (1).

Geography
General Level 2005 (continued)

8. (a) *1 mark for correctly placing points on graph. 1 mark for completing line appropriately on graph.*

   (b) **Pie chart** (1)
   Possible reasons could include:
   - good at showing percentages (1)
   - can be coloured to emphasise differences (1)
   - allows comparisons between countries (1).

   Divided bar graph (1).
   - All of the above reasons apply.
   - Also easier to draw than pie charts (1).
   - Bar graph is good at showing differences (1).

   Rank Order (1)
   - Lets you see where most refugees come from (1).

9. *1 mark per valid point, 2 marks for a developed statement.*
   - e.g. American farmers have much better farming equipment than the Mexican farmers so can produce much greater quantities of maize (2).
   - The US government gives subsidies to their farmers (1).
   - Family farms in Mexico cannot compete with big American commercial companies (1).
   - Cheap maize being sold in Mexico makes it more difficult for Mexican farmers to sell their maize (1) and so they get poorer (1).

10. *Award marks for reasons given. Accept answers which only partially agree. 1 mark per valid point, 2 marks for a developed statement.*
   Most likely answer would be:
   Short Term Aid
   - So many people were made homeless that they would need to have emergency shelters (1).
   - Water supplies would have been contaminated so clean water would be essential (1).
   - To help stop further death and disease (1).
   - People would need food because their crops would have been destroyed (1) and medicines would be required for people made ill by dirty water or by starvation (1).

   Accept valid reasons for Long Term Aid.

Geography
Credit Level 2005

1. (a) *1 mark per valid point; 2 marks per expanded point. Both river and valley must be referred to for full marks. Max (1) for GR.*
   Possible answers might include:
   R Carron is in its lower course (1) flows in easterly direction (1) has many meanders e.g. 894826 (1) is tidal up to 878823 (1) channel is wide (1) gets slightly wider near to Carron House (1).
   River Carron has wide flat flood plain (1) valley floor is much narrower at Lochlands (1) contours are very near to river (1) ox bow lake on valley floor at 868812 (1) valley sides climb gradually to N & S (1) waterfall (1)

   (b) *No marks for choice. 1 mark per valid reason; 2 marks per expanded point. Accept negative points against alternatives. Max (1) for GR. Accept list.*
   Possible answers might include:

   Industrial:
   - canal connecting Forth & Clyde (1)
   - may be used to carry goods (1)
   - several 'works' (1) e.g. 879825 (1)
   - plenty of residential areas e.g. Carmuirs (1)
   - near works which would provide housing for workers (1)
   - has good A road & motorway network (1).

   Service Centre:
   - has numerous services for residents (1)
   - churches, railway & bus stations in CBD (1)
   - roads connect Falkirk easily to motorway (1)
   - plenty schools for local children (1) e.g. 882807 (1).

   Tourism & Recreation:
   - canal probably used for tourist craft nowadays (1)
   - tourist info centre of Falkirk (1)
   - historical interest with Roman Forts etc. (1) 843798 (1)
   - & Antonine Wall (1)
   - golf courses found on floodplain area (1)
   - sports centres also in built up area (1) e.g. 866806 (1)
   - Falkirk Wheel is tourist attraction (1)
   - tourists could visit museum at Callendar House (1) 898794 (1).

   (c) *1 mark per valid point; 2 marks per expanded point. Mark 1:3 2:2 3:1*
   Possible answers might include:

   ADVANTAGES:
   - near market to sell produce (1)
   - good road connections (1) via B902 ' A88 ' motorway (1)
   - flat land easy to use for growing crops (1)
   - machinery (1)
   - water from stream available for livestock (1)
   - not too near river to avoid flood danger (1)

1. (*c*) (continued)

DISADVANTAGES:

- too near residential area disturbance by people walking over fields (1)
- near lots of industry – might affect crops adversely (1)
- land flat – may be liable to flooding (1)
- relatively inaccessible (1)
- only B road nearby (1)
- land under pressure from urban expansion (1).

(*d*) *1 mark per valid reason; 2 marks per expanded reason. Mark 1:3 2:2 3:1*
Possible answers might include:
8781: land use farmland (1) flood plain, therefore unsuitable for building (1) could be marshy (1) probable flood danger (1)
8777: land use farmland – probably hill sheep (1) higher up so less suitable for residential use – colder, not sheltered (1) steep slopes unsuitable for building (1) marshland (1).

(*e*) *1 mark per valid explanation; 2 marks per expanded point.*
Possible answers might include:
- in area open land with room to expand (1)
- near motorway link to north A road to south (1)
- with free flowing traffic due to roundabouts (1)
- fairly level land – good for building (1)
- near to Larbert/Falkirk for employees (1)
- nearby golf course for employee recreation (1).

(*f*) *1 mark per valid point; 2 marks per expanded point. No marks for grid references. Accept negative points for alternative grid square.*
Possible answers might include:
YES:
- roads converge (1)
- ring road feature (1)
- tourist info (1)
- railway station (1)
- churches (1)
- town hall (1)
- other public buildings (1)

NO:
- not in centre of built up area (1)
- bus station in 8979 (1)
- lots of housing in this square (1)
- schools not usually in CBD (1)
- more roads meeting/ring road to south of grid square (1).

2. *1 mark for each valid point. 2 marks for a developed point. Marks should be given for relevant points made in diagrams.*
Created by glacial erosion (1) through abrasion and or plucking (1). As the glacier moved down the valley, rock fragments became frozen in the base of the glacier (1) and were plucked away as the ice moved on (1). Ground moraine scraping over the land surface is known as abrasion (1). As a result the valley became deeper, straighter and wider (1).
Or any other relevant explanatory points.

3. (*a*) *1 mark for each valid point. 2 marks for developed point.*
Possible answers might include:

Forestry/Walkers
- Dangerous machinery moving along access roads (1)
- areas of forest fenced off (1)
- felling operations closing down access paths (1)
- re-routing walkers causes time to be added to journey (1)
- coming off the hill not knowing about felling in progress is dangerous (1)

Residents/Busy Road
- Resident community would object to increased traffic on roads (1)
- lengthens journey time to main road and beyond (1)
- loss of privacy (1)
- increase in noise levels and pollution from vehicles (2)
- increase in litter (1).

Hillwalkers/Areas of Scientific Interest
- Foot path erosion in areas of scientific interest (1)
- damaging fragile plants and ecosystems (1)
- highland takes long time to recover (1)
- so difficult to sustain research (1)
- increased costs for land owner in footpath repair and maintenance (1).

(*b*) *Mark 2:4 or 3:3. Up to 3 marks for justification of one technique*
Possible Techniques:
- Questionnaire given to local community (1)
- or people involved in recreation (walkers, cyclists) (1).
- Photographs of area (1)
- counts of walkers, cyclists, canoeists (1)
- Field sketches (1)
- annotating series of maps (1).

Geography
Credit Level 2005 (continued)

3. *(b)* (continued)

Justification
- Would allow students to identify conflicts/problems (1)
- could see when peak periods of usage occur – weekly/annually (1)
- speaking to specialists e.g Forest Manager will allow them to gain accurate information about damage (1)
- other organisations like Ranger Service will also be able to give detailed information (1)
- by asking local residents and chalet owners students can gauge opinions from people in area (1).

4. *(a)1 mark per valid comparison.*
Possible answers might include:
- Wind is much stronger in North Scotland, 35 knots compared to 10 knots in SE (1)
- temperatures are also lower by 3°C (1).
- Snowing in N Scotland but raining in SE (1)
- cloud cover heavier in N Scotland, 8 oktas compared to 4 oktas in SE (1)
- wind in N Scotland from different direction NW compared to N in SE (1).

(b)1 mark per valid point; 2 marks per expanded point. Accept both sides of argument.
Possible answers might include:
YES:
- winds light in high pressure areas in SE (1)
- drizzle and heavy cloud cover in S Ireland linked to cold front passing (1)
- cloud and stronger winds in NE England connects to cold front (1)
- effects of cold air still in Scotland (1)
- NW winds in Scotland match the isobar pattern.
NO:
- wind too strong in N Scotland isobars are far apart rather than close together (1)
- should be raining in middle of England where cold front is (1)
- too much cloud in SE where high pressure in Q4B (1)
- drizzle in S Ireland usually after warm front (1).

5. *1 mark per valid point. 2 marks for a developed point.*
- Unemployment (1)
- will lead to increased pressure on social services (1)
- social security payment will increase (1)
- local shops may close due to local people having little money to spend (1).
- House prices will fall as people leave the area in search of work (1).
- Small local businesses that provide services to the steelworks may also close (1).
- Increase in crime (1)
- general breakdown of infrastructure (1)
- area becomes run down (1).

6. *1 mark per valid point. 2 marks for a developed point. Mark 2:4, 3:3 or 4:2.*
Advantages:
- Larger fields enable machinery to be used (1)
- making work quicker and easier (1).
- Removing hedgerows gives more land for farming and bigger profits (1).
- Larger farms mean that economies of scale can be practised (1).
- Having fewer farm workers means that the wages can be higher (1).
Disadvantages:
- Destruction of hedgerows and trees causes loss of wildlife (1).
- Reduced number of farm workers increases unemployment (1)
- and forces farm workers to leave the countryside (1).
- Consequently the number of services in the countryside declines (1)
- e.g. schools, village shops etc (1).

7. *1 mark per valid point. 2 marks for a developed point. Marks awarded for 1) Differences 2) Explanation*
Answers may include:
- Area A appears to be a new housing area because streets are short and varied in layout (1) which is a sign of modern attention to an attractive environment (1).
- Many streets are cul-de-sacs and there are no through roads for safety (1); there is a lot of open space suggesting cheaper land on the edge of town (1) so the environment is likely to be less noisy, polluted and congested than that of the inner city (1).
- Area B looks to be an older area because streets are longer and more regularly laid out with little attention to visual appeal (1) many of the roads can be used by through traffic, meaning more noise and pollution and less safety (1) open space is mostly occupied by works and railways (1) suggesting industry close to housing – a feature of older, inner-city areas (1). In Scottish cities like Glasgow, building types in such areas tend to be tenements (1) with high population densities and with less room for gardens/green open space (1).

8. *(a)1 mark per valid point, 2 for a developed statement.*
e.g. Some countries are too dependent on just one or two exports (1). For example, Ghana depends on cocoa for 80% of its exports (1) and so is vulnerable if the world price for cocoa drops (1). This would mean that cocoa farmers would receive less income and so have less to spend, affecting other businesses in Ghana (2). These countries would make more money if they were able to process some of their goods instead of just exporting mainly raw materials (2). Or they would be less affected by changing world prices if they had a greater range of goods to export (1).

8. (b) *Mark as 2:4, 3:3 or 4:2 for choice of techniques and justifications.*

Pie charts (1):
these would be suitable because the figures are expressed as percentages (1) and a series of charts could be drawn, one for each country (1). This would allow easy comparison (1) and colour could be used to highlight the different segments (1).

Bar graphs (1)/divided bar graphs (1):
a divided bar graph could be drawn for each country and they could be placed in rank order (1) so that it would be easy to see at a glance which countries depended the most on just one or two commodities (1). A histogram would also show this in the same way; with the highest percentages drawn first (1).

Tabulating (1):
a table could be drawn up with the countries placed in order, so that the most dependent countries appeared first (1). The table could be further subdivided into categories to show groups of countries over 90%, between 80-90% etc (1) and even colour coded to highlight these differences (1).

Or any other valid technique.

9. *1 mark per valid point. 2 marks for a developed point.*
Answers may include:
- In Ethiopia birth rates are high because of little access to education and family planning (1) whereas the United Kingdom these are widely available (1).
- In Ethiopia birth rates are high because young people are needed to support old age/illness (1) whereas in the United Kingdom health and pensions provision make this unnecessary (1).
- In Ethiopia birth rates are high because young people can earn wages from working – adding to family incomes (1) whereas in the United Kingdom education is compulsory to 16 years (1).
- In Ethiopia birth rates are high because females have little access to higher education/careers (1). In the United Kingdom education is available and many females put off having children (1).
- Any other relevant point.

Geography
General Level 2006

1. (a) *No marks for grid reference.*
Annotated diagram can get full marks.
There is a very pronounced meander in 7161(1); the neck is very narrow (1); the outsides of two bends are almost touching (1); continued erosion on the outside of these bends could allow the river to break through (1); especially when water levels are high or there is flooding (1); the river is then likely to by-pass the loop (1); and follow a straighter course (1); leaving the cut off loop as an ox bow lake (1); separated from the new channel by deposition (1).

(b) *Maximum one mark for relevant grid reference.*
Mark 3:1, 2:2, 1:3

Advantages
In the northern part of its route the footpath affords good views of the river valley (1) especially crossing the bridge (725670) (2); from Nashenden Farm (730660) the path climbs to the top of the downs via a fairly gentle gradient on a spur (1); for much of its length (eg in 7263 and in 7859) the path follows the ridge, so the route is fairly flat for walking (1); from the top of the ridge good views are possible over the surrounding countryside (1); the countryside is varied and interesting, including much woodland (eg in 7263) (1); there are several public houses on the route (eg at 734626) (1) which may offer food and/or accommodation (1); there is a camp site fairly nearby (at 747638) (1); there are various sites of historic interest, including the tumulus (at 727653) and White Horse Stone (750601) (2).

Disadvantages
Not pretty sites for walking (1) too near motorways (1) Quarry would be unattractive (1) views spoiled by trees (1) industrial estate nearby at 7359 (1) would be better to follow river closely (1).

(c) In 7559 the most likely use is shelter belt (1), or ornamental row of trees (1), to screen the railway (1).

In 7859 trees prevent soil erosion on the steep slope (1), or may be the only way the farmer can use the steep slope commercially (1), natural woodland (1).

In 7968 – orchard (1).

One mark relating to each square.

(d) *No mark for choice.*
Comparative points accepted.
No mark for grid reference.

Possible answers might include:

Geography
General Level 2006 (continued)

1. (*d*) (continued)

X: The land is flat for building on (1); the area is well-drained and there's no danger of flooding (1); there are already settlements nearby, so development is not using up unspoiled countryside (1), and there are likely to be services available already (1), including schools reasonably near (eg at 805658) (1); there are many leisure facilities in the area (1), including country park and ski centre (1), space for building (1).

Y: Nearby power supply (1); close to motorway junction (1); existing services in village (1); brownfield site (1); scenic area, with walks nearby (1).

(*e*) *One mark per valid point – 2 marks for expanded point.*

Possible answers might include:

The land is flat – very few contours – so easy to build on (1); large areas of open space on site for storing raw materials (1); the River Medway and the lakes/reservoirs provide a water supply (1); there are numerous woodland areas fairly near for raw materials (1); there are good road communications for cheap transport including a motorway and the A228 and A229, both principal routes (2); the railway is very useful for moving bulky goods (1); there is housing beside the mills for a workforce (1).

Must use map evidence.

(*f*) *One mark per valid point, 2 marks for expanded point.*
Maximum one mark for grid references. Mark 3:1, 2:2 or 1:3.

Possible answers might include:

Benefits could include: the piers (eg 780698) (1) and Historic Dockyard suggest that the river helped to create employment (1); they also suggest trade which brings/brought wealth to the area (1); the river valley acted as a routeway (1) which also made the settlements a focal point for trade (1); nowadays the Adventure Park (751699) and the Marina and Leisure Park (7869) all suggest river-based recreation (2); the Castle (739685) and the Fort (760683), built to guard the river mouth are now tourist attractions and bring money to the area (2).

Problems include: the river was a barrier to east – west communications (1) meaning road and railway bridges and a tunnel (1) had to be built (1) and would be expensive (1); present volumes of traffic using a limited number of routes across the river (1) are likely to cause congestion (1) leading to inconvenience and pollution (1); some land beside the river is flat and liable to flood (1) so couldn't be built on (eg Temple Marsh 7367) (1); the river's shallow

1. (*f*) (continued)

water, mud and winding course (eg 7468 and 7568) might make navigation difficult for large, modern ships (1) and so lead to a loss of trade and industry (1).

2. (*a*) *3 marks for all four correct, 2 marks for two or three correct, 1 mark for one correct.*

drumlin	B
terminal moraine	C
outwash plain	D
boulder clay	A

(*b*) *Mark 3:1, 2:2, or 1:3. One mark per valid point.*

Possible answers might include:

Arable Farming: is found on the boulder clay because it's more fertile (1) and as it is mainly flat land it's easy for machines to work on (1).

Forestry: is more suited to the terminal moraine because the slopes make the machinery difficult to use (1); trees can grow well here though the crops couldn't (1) because it's less fertile (1).

Quarrying: takes place on the outwash plain because it isn't fertile enough for many crops (1) and the sand and gravel can be used in the building industry (1).

3. *1 mark for each valid point, 2 marks for a developed point.*
For full marks answers should refer to both Dunkeld and Shetland.

Possible answers might include:

Shetland is in an anticyclone high pressure area (1), which means skies will be clear (1), allowing the sun to shine in Shetland (1).

Dunkeld is affected by an occluded front (1) and will bring heavy rain (1).

Wet because of the front (1).

4. (*a*) *One mark for each point E and F correctly plotted.*

(*b*) *Accept Yes/No answers.*

Yes: as sunshine hours increase, so does temperature (1), this shows a positive link (1).

No: point F has low hours of sunshine but a high temperature (1).

Details of temperature or sunshine (1).

5. *Answers can refer to either part of the statement. One mark for a valid point, two marks for a developed point.*
Accept Yes/No answers.

Yes: the North Sea is becoming polluted (1). There is the possibility of leaks from drilling rigs or pipelines (1), or oil spills from tankers (1), which kill sea birds (1). Fertilisers from farmland are carried by rivers into the sea (1).

5. continued

Industrial fishing can damage fish stocks (1), and ruin fishing for the future (1). Dumping of sewage and industrial waste also pollutes the sea (1).

No: we need to catch fish to feed the large population of W. Europe (1). Oil and gas are essential for industry (1) and to maintain our high standard of living (1). Farmers need to use fertilizers to produce good yields (1). It is more efficient to fish on an industrial scale (1) Spawning areas for fish are long way from industrial parts (1)
… or any other valid point.

6. *One mark per valid point, two marks per developed point.*

Possible explanations might include:

Farmers have lost money because of subsidies being cut/price cuts (1), so different uses would help to replace money lost (1). Holiday cottages can be made from cottages no longer needed for workers (1) because of mechanisation (1), and can be let out especially in summer (1). Encouraging farm visits through kids' zoos, etc will bring in money (1) and help people to care for the country side (1). Credit reference to points not made in Q6A.

7. (a) *No mark for choice. Accept Yes/No answers.*

If "Yes" is chosen, answers could include:

Government Aid allows rate-free periods (1); areas chosen by the Government for assistance are likely to have labour available because of unemployment/closure of older industries (1), and many workers in these areas may have relevant skills/experience (1); in addition grants may be available to train workers, further cutting costs (2); by developing good transport links (1), and the provision of all services including power and water (1).

"No" answers may include:

Pleasant environment more important for workers (1) if factory is large single storey, lots of flat land is necessary (1) nearby markets is an important factor (1) transport for raw materials also important (1).

(b) *Mark 2 for techniques and 2 for reasons.*

Possible answers include:

Visit the site and note which firms have premises (1); this would give information about the nature of the estate (1) eg does it specialise in high-tech firms? (1)

Interview owner/manager of firm(s) (1); this will make it possible to identify exact importance of the different location factors/reasons for choice of site (1).

7. (b) (continued)

Questionnaire to workers would allow information to be gathered about: area where workers come from (1) or how they travel to work (1) or their opinion of this as a workplace (1) if enough (minimum thirty) questionnaires are issued this should give a representative view of workers' opinions (1) or if questions are kept few and easy to answer (eg tick box) then this should encourage a good response (1).

Fieldwork would allow an environmental quality index to be compiled, based on specific criteria (1); this would enable comparisons to be made with other areas (1).

8. Set of population data B (1). Credit rest of answer if mistaken choice.

Possible reasons might include:

ELDCs have a large number of children, so pyramid will have a wide base (2).

Death rates are high, so pyramid will taper sharply at the top (2). Living conditions are poor so life expectancy is low (2). Population of ELDCs mainly work in agriculture, so a high percentage of the population lives in the countryside (2). Fewer industrial or service jobs, so low urban population (2).

9. (a) *Accept Yes/No answers. Maximum 1 if straight list of figures.*
1 mark per valid point.

If "Yes": Japan is a rich industrialised country because it exports modern machinery (1), 64% (1). The Japanese are able to produce a range of high quality goods (1). The Japanese import large amounts of cheap raw materials (1), and export more expensive manufactured goods.

If "No": Japan cannot grow enough of its own food (1). Unlike USA/EU, Japan needs to import oil to convert to petro chemicals (1) or as a source of power (1). Japan needs to import a lot of manufactured goods (1) 43% (1).

Or any other valid point.

(b) *A maximum of two marks for valid techniques.*

Possible techniques include:

Bar graphs
Divided bar graph
Table
Pictograph
Flow charts

Possible reasons:

Eg bar graphs give a good visual comparison of amounts (1); less calculation needed than for a pie chart (1); bar charts will highlight smaller amounts than a pie chart (1).

Geography
General Level 2006 (continued)

9. (*b*) (continued)

Eg divided bar graph is good for showing percentages/shares of a total (1); similarities or differences can be emphasised using different colours/shading for different parts of a total (1). The divided bar graph takes less time to process than a pie chart (1).

Don't accept repeat of reasons.

10. *No mark for choice. Mark 3:1, 2:2 or 1:3.*

Answers might include:

Eg: Improved water supply; clean water supply will reduce disease (1); more water for irrigation will increase food supply (1); people no longer have to carry water long distances (1).

Eg: Education opportunities; more teachers means more children can attend school (1); literacy rates will improve (1); colleges develop a skilled workforce (1) and this should help attract industry (1).

Geography
Credit Level 2006

1. (*a*) *1 mark for each correct answer:*

    A   -   Beinn na Caillich
    B   -   Allt Nathrach
    C   -   Kinlochleven

(*b*) (i) *3 marks for all four correct, 2 marks for three or two correct, 1 mark for one correct.*

    arete   057563
    hanging valley   165553
    truncated spur   201556
    corrie   197584

(ii) *1 mark per valid point. Fully annotated diagrams may gain full marks. Credit references to frost shattering, abrasion, ice plucking and interlocking spurs.*

Truncated spur – is formed when the slope of a hill is eroded by a glacier (1); as the ice moves down the valley it abrades the sides of hills (1); and ice at the edge of the glacier freezes on to the rock and plucks it away (1). When the ice melts, the slope is left as the steep side of a U-shaped valley (1) and may have crags or cliffs where erosion was greatest (1).

(*c*) *No marks for choice, 1 mark for each valid supporting reason, 2 marks for developed points. Accept yes/no answers. Maximum 1 mark for grid references.*

Possible answers might include:

Yes – it is a good use for an old industrial site as the buildings will not be left derelict (1) and there may be some jobs for the local community (1); it is a good location because all of these sports can be done on the surrounding mountains (1) while the river and loch could also be used for water sports (1); forests such as those in 1762 (1) could be used for orienteering (1); a long distance footpath (West Highland Way) runs through the area and so there are likely to be lots of people interested in the centre (1); mountain rescue teams could use the centre to help with their training (1).

No – there is only one B class road so access is not good and it is a long drive from the nearest main road or the nearest towns (2); the surrounding area is ideal for these sports, so it would be better to have the centre in a place further away, so that people could prepare before they visit mountainous areas (2); Kinlochleven is only a small settlement so there will be few customers and there may not be a big enough pool of labour (2).

1. *(d)* *1 mark per valid point, 2 for developed statements. Maximum 1 mark for simple statements linking land use to landscape.*

Possible answers might include:

the settlement of Glencoe village is on flat land (1) suitable for building (1), it is beside the River Coe and is located here because this is where roads meet (1). The mast is on the hilltop where it receives good reception (1). Forests are found on steep slopes where the soils may be thin (1); sheep are found on steep slopes or high land because the ground is unsuitable for crops (1); there are hotels, caravan and campsites and footpaths to cater for the many tourists visiting the area for the spectacular scenery (2).

(e) *No marks for choice, 1 mark for each valid supporting reason, 2 marks for developed points. Mark out of two if the answer is general and does not refer to the map. Accept yes/no answers. Credit references to helping Scotland reach its target for renewable energy and/or lack of air pollution etc.*

Possible answers might include:

Yes – the land is very high and exposed (1011 metres) so it will be a good site to catch the wind (1); there is a main road (A82) close to the site, so access will be easier (1); there are very few houses for miles around so there will be less protests from local residents (1); the turbines could be placed below the summit so there would be less visual intrusion (1); there are no trees which would need to be cut down for the turbines or to create access (1).

Or any other valid point.

No – the land belongs to the National Trust who would be unlikely to let this happen (1); wind turbines on Buachaille Etive Mor would spoil the view from the main road for passing motorists and it would also spoil the scenery in one of Scotland's most popular tourist areas (2); the slopes of the mountain are far too steep and it would be impossible to get heavy machinery on to the top (2); as it is far away from any big settlements it would be expensive to transmit electricity from Buachaille Etive Mor (1); the area is obviously popular for walking as there are car parks (eg 213560) and footpaths (eg 218550) and building a windfarm would cause conflict with hill walkers (2).

Or any other valid point.

2. *Marks are awarded for noting and explaining the differences. Maximum of one mark for description of differences.*

Possible answers might include:

wind direction at Stockholm is NW whereas at Belfast it is SW (1 mark for description), due to the different alignment of the isobars (1) and the fact that winds circulate in an anticlockwise direction around a depression but clockwise around a centre of high pressure (1). In Stockholm it is dry but in Belfast it is wet (1 descriptive mark) because Stockholm is in a ridge of high pressure whereas Belfast is in a depression (1).

There are 8 oktas of cloud cover in Belfast because it is close to the warm front, whereas Stockholm is not yet affected by the clouds associated with the advancing warm front (1). Stockholm is in a ridge of high pressure, so is experiencing dry conditions, unlike Belfast which is close to the warm front/in the warm sector of a depression (1)

Temperatures in Belfast are warmer than in Stockholm as it is in the warm sector of a depression whereas Stockholm is in the cold sector (1).

3. *1 mark per valid point, 2 for developed statements. Maximum of one mark for basic references to land use or a list.*

Possible answers might include:

new roads such as the Trans Amazon Highway (1) are being built to open up the interior (1); trees are being cleared to make way for ranching which gives bigger profits (1); parts of the forest have been flooded by HEP (1) schemes to provide water and electricity (1); logging also destroys the forest (1) but governments can charge companies for logging rights in the rainforests (1) and so bring in money to help with their rapidly expanding populations (1); new settlers are destroying the edges of the rainforest through slash and burn agriculture (1); while mining companies have destroyed areas of forest in order to extract minerals (1) which can be sold abroad and earn much needed foreign currency for the ELDCs (1).

4. *(a)* *No marks for choice; 1 mark per valid point, 2 for developed statements. Accept yes/no answers.*

Possible answers may include:

Yes – organic food being grown is better for the environment (1) because no chemical fertilisers or pesticides are used (1); mechanisation means that the countryside becomes more productive (1) as the farms are more efficient and produce higher yields (1); set aside land might be allowed between fields to help create a more natural environment (1); the green veins idea would encourage more wildlife (1) and perhaps more ponds and wetlands for birds (1).

Geography
Credit Level 2006 (continued)

4. (*a*) (continued)

No – set aside fields look messy (1) removing hedges spoils the prettiness of the countryside (1) and harms wildlife as their homes/habitats are destroyed (1); mechanisation causes more noise and air pollution (1) and causes job losses which lead to people moving away (1); this can also result in loss of local services/school closures etc (1).

(*b*) *Maximum of two marks for reference to any one technique. No credit for same justification for different techniques.*

Possible Techniques:

Bar graph (1), pie chart (1), land use map (1).

Possible Explanations:

Bar graph: separate bars would clearly show the amount of land for each category (1); and could be coloured to emphasise the differences (1).

Pie chart: could change the figures into percentages (1) and show the proportion of land used for each crop (1)

Land use map: would show where the crops are grown (1) colour would show the most abundant crops (1); patterns of land use could be identified such as high intensity land uses close to the farmhouse (1).

5. (*a*) *No marks for description; one mark per valid point, two for developed statements.*

Possible answers may include:

Inner City area: were built close to the original centre of the settlement, near to services, business and industry (1); grid-iron street patterns were used in the inner city to make the best use of space (1); houses are closer to industry in older areas because they were built when people had to walk to work (1); tenement/terraced housing is found in inner cities because this allowed a very large number of people to live in a small area close to industry (1); expensive land in the crowded inner city allowed little room for gardens or open spaces (1); inner city environmental problems are often a result of derelict industrial sites, run-down older housing and traffic congestion(1).

Suburbs area: new housing is further away from the centre because recent development has spread outwards into areas which were not already built up (1); modern planners have used cul-de-sacs to discourage through traffic as a safety measure (1); also a more varied road layout is used to create a more varied/less boring environment (1); new housing can be built further away from the city as people can now commute by car to their work place (1); modern housing allows lower densities of population and room for bigger gardens because of cheaper land

5. (*a*) continued

on the outskirts (2); this helps create a more pleasant environment where industry and housing areas are usually separated (1).

(*b*) *At least two techniques must be described. Maximum of three marks if no reasons given, or if reference is made to only one technique. Mark 2:3 3:2.*

Possible answers might include:

Fieldwork based on a land use survey (1) could be used to produce an environmental quality index for each area (1), using the same criteria for each, such as evidence of dereliction, litter, quality of open space and number of empty properties (1); reasons might be that this would allow a comparison to be made (1) based on the use of facts rather than just opinion (1); provides accurate, up-to-date information (1). Photographs of the areas could be taken and displayed side by side (1); these would capture the appearance of the town in greater detail (1); map studies using a variety of maps (1) would show up differences in the amount of open space and/or land use (1); this would allow even widely separated areas to be compared without the need for travel costs/time (1).

6. (*a*) *One mark per comparative point.*

Possible answers might include:

Tokyo will have increased from about 17 million in 1970 to 30 million in 2015 whereas Jakarta has gone from 3 million up to 22 million (1); so Jakarta's population has increased by 6 million more people (1); since Tokyo started off with a lot more people than Jakarta in 1970, it is Jakarta that has gone up more in percentage terms (1); in Tokyo the population will have nearly doubled whereas in Jakarta the increase will be over 700% (1). Tokyo's rate of increase is quite slow between 1994 and 2015 while Jakarta is still increasing rapidly between these dates (1).

(*b*) *Accept yes/no answers.*

Answers may include:

No: Jakarta is in a poorer country so more strain on resources from rapid population growth (1); economic development cannot keep up with population growth (1); this could lead to lack of food (1) illness/disease because people have a much lower standard of living than in Tokyo (1); there will be a strain on education and health services (1) and a lack of housing (1) causing sprawling shanty town developments (1); Indonesia may need to import more than they export and go into debt (1) as they borrow money from EMDCs and have to pay high interest rates on their loans (1).

Yes: there will be problems of overcrowding in Tokyo (1) which could lead to further environmental difficulty such as increased

6. *(b)* continued

pollution from vehicles as people commute to work (1); there may be a lack of suitable jobs causing people to become destitute (1) and forcing some to live on the streets as happens in many of the world's major cities (1); there will be a lack of suitable land for building in Tokyo (1) as Japan is already a very densely populated country (1); valuable farmland may have to be sacrificed to build homes for all the extra people (1).

Accept any other valid points.

7. *No marks for choice. Accept answers which refer to both physical and human factors.*

Possible answers might include:

Physical factors:

Relief: flat land is easier to build on (1) and also to farm because soils are usually more fertile (1) transport is easier in areas of flat land (1) and so more people choose to live in these areas giving higher population density (1).

Natural resources: if an area has lots of natural resources such as water, wood, oil, coal, people will want to live there (1) as there will be more job opportunities (1) trade links and industrial development (1) which in turn can create prosperity (1).

Human factors:

Government decisions: if the government gives grants or some kind of financial incentive, they can attract employment into an area (1) which leads to better incomes and higher prosperity (1) which will result in more people wanting to live in the area (1).

Employment opportunities: in a poor area there will be few jobs (1); if it is inaccessible not many people will live in these areas (1); but if the area is easily reached and has good transport links there is likely to be more work available (1).

8. *One mark per valid point, two for developed statements.*
Mark 3:3, 4:2, 2:4.
Accept 1 lift for advantages and 1 lift for disadvantages.

Possible answers may include:

Advantages: the electricity generated will help save China money (1) and this will go towards paying for the dam (1); 10 million people will be safe from the danger of flooding (1); their homes and crops will no longer be at risk (1); HEP is a clean source of power which will reduce pollution at local and global level (2). The new economic development will encourage more foreign investment (1); this will improve the standard of living for millions of Chinese by creating new job opportunities (2); trade links with rich countries can help the Chinese economy to grow (1). If

8. continued

China's economy is booming, benefits trading partners due to more imports (1)

Disadvantages: China could fall into debt (1) which will become a burden on the Chinese economy (1) and cause it to buy less from trading partners as it could end up paying huge amounts of interest on its loans to foreign banks (1); displaced people have had their way of life destroyed (1); many places of historical and archaeological value will be lost forever (1) the dam could be destroyed by an earthquake (1); this could lead to deaths of millions of people due to flooding (1).

Geography
General Level 2007

1. (a) *3 marks for four correct, 2 for three or two correct, 1 for one correct.*

Physical Features	Grid Square
Steep southwest facing slopes	6286
Flat land	7490
Broad ridge running East-West	6586
V-shaped valley	6193

3ES

(b) *One mark per valid point. For full marks candidates must mention river and valley features. Maximum one mark for grid references.*
To begin with the river Frome has small meanders (1) and a narrow flood plain (1). As the river flows south-east (1), the flood plain gets wider (1). There is a confluence with another river at 692913 (2). The south-west slopes are steep (1) and rise about 100 metres above the valley bottom (1).
Any other valid point. (1)
4KU

(c) *Mark 2 marks for techniques and 2 marks for reasons.*
Measure the width (1). Measure the depth (1). Measure the speed (1). Draw a field sketch (1). Take a photograph (1). Use a flow meter to record speed (1). Use an orange to measure speed (1). Oranges are cheap and readily available (1). Oranges are clearly visible (1). You could take photographs and compare the features of the river (1). Use a measuring pole to help find if the channel in the river was shallow on the inner bend and deeper on the outside (1). Use a float over a set distance and time to measure the speed of the river (1). Flow meters give more accurate readings (1).
Any other valid techniques and reasons.
4ES

(d) *1 mark per valid statement. 2 marks for a developed point. Maximum 1 mark for grid references.*
Answers might include;
For: It is near the A35 road which would give the park easy access for visitors (1). The woodlands have many paths for walking and cycling (1). You could go birdwatching (1) or visit Hardys Cottage (1). There are car parking facilities (1) at GR 725922 (1).

Against: The road splits the park in two making it dangerous to cross (1). Large numbers of visitors could cause footpath erosion (1) especially near archaeological sites (1). These sites could be vandalised if the area became more popular (1).
4ES

1. (continued)

(e) *1 mark per valid statement. 2 marks for a developed point. No marks for grid references.*

The land to the west is relatively flat (1). This makes building roads, houses and factories easier (1). The area has good road access (1). There is a river to the north (1) and you would need to put in a new bridge (1) and this would cost a lot of money (1). Flood defence schemes are very expensive (1) and there will be a danger of flooding (1): homes would be damaged by floodwater if they were built there (1).
3ES

(f) *No marks for choice. 1 mark per valid point, 2 for developed statements. (No marks for grid references).*
Tourist resort: There are good facilities for outdoor activities such as the National Cycle route (1) and River Frome for water-based activities (1). There are also many features of historical interest (1) such as the remains of a Roman temple at Maiden Castle in square 6888 and Maumbury Rings at 6989 (1). There are at least three different museums for tourists to visit (1).

Market town: The area is very accessible (1). There is a showground marked in 6991 and 7091 (1) which indicates that there is probably an annual agricultural show (1). There are many farms in the surrounding area (1) such as Herringstom Farm in 6887 (1) and these farms will send their produce in to Dorchester on the good network of roads which lead into town (1). The land around Dorchester appears to be well drained and much of it is gently sloping, so it will be good for different types of farming (2).
3ES

(g) *1 mark per valid statement. 2 marks for a developed point.*
Answers might include:
The farm is too close to the main road and the busy traffic for slow-moving farm vehicles (1). There will be air pollution from the main road (1). It is close to housing estates from where people might trespass and trample crops (1). People might leave gates open and drop litter (1). The noise of traffic and people may disturb the animals (1). It is difficult to use machinery on the steep slopes to the south-west (1).
Any other valid point.
3ES

2. *A well annotated diagram could score full marks.*
The glacier pushes material in front of it (1) and when the ice melts, material carried by the ice is deposited (1). This material forms a ridge (1): a mixture of rocks, stones and debris (1).
Any other valid point.
3KU

3. (a) Air mass B is Polar Continental (1)
Air mass D is Tropical Maritime (1)
Mark 1:3, 2:2, 3:1
2KU

(b) **BENEFITS include:** Plenty of sun to ripen crops (1), enjoyable weather for beach holidays (1) and good levels of income for the tourist businesses in the resorts (1); less fuel and fewer power supplies needed, saving people money (1), and less pollution from power stations (1).

PROBLEMS include: Shortage of water may cause a need for expensive irrigation of crops (1). Drought can cause water shortages (1) and restrictions on use, eg hosepipe bans (1). Likelihood of damage caused by forest fires (1). Any other valid points.
4KU

4. (a) *1 mark per valid description. Answers must refer to temperature and rainfall for full marks. Answers must refer to figures for detail.*
Answers might include:
Temperatures are high all year (1) – above 25 degrees centigrade (1). The highest temperature is 29 degrees in July, August (1). There is a small range of temperature (1). There is rainfall all year (1). Highest rainfall is 260 millimetres in December (1) and the lowest is 175mm in July (1). There are no seasons (1) as it is hot and wet all year (1).
3ES

(b) *No marks for choice: 1 mark for a simple point and 2 for a developed point. (Note that Yes/No answers are acceptable.)*
Agree: New roads have made the area more accessible and easier to get around (1). This brings in new jobs as commercial farming, mining and industry develop (1). New dams reduce the risk of flooding (1) and provide water for a range of activities eg farming industry and recreation (1). Forests have been cleared for cattle ranching, increasing the food supply (1). All of the developments can help the country to develop trade with others and improve the standard of living of the local people (2).

Disagree: Areas of forest have been cleared to make way for new developments, reducing the area which provides food for the locals (1). They have been forced out of their homes (1). There is more soil erosion as there are fewer trees to hold the soil (2) so fewer food crops can be grown (1). New developments have attracted outsiders who have brought diseases (1). Local people feel that their traditional way of life is disappearing (1). Mining scars the landscape and causes pollution (1). Most of the food produced on the farms is exported (1).
4ES

5. (a) *1 mark per valid point. 2 marks for a developed statement. (No marks for description).*
Answers may include:
Milk quotas could be why the farmer has a smaller dairy herd now (1). He may also have received grants from the government or EU to restore wetlands (1) and also to plant more trees on his land (1) there will be less need for farm workers due to mechanisation and so the farmer has converted their old houses into holiday cottages to gain extra income (2) organic crops such as potatoes are grown as there is a bigger market for them now (1).
4KU

(b) *3 marks to be awarded as follows:*
  • Drawing two lines accurately to further subdivide the graph for organic potatoes, set-aside and restored wetland (2)
  • Shading each of the three sections appropriately (1)

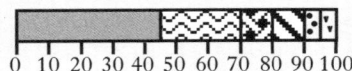

3KU

6. *1 mark per valid point. 2 marks for a developed statement.*
No: The cement works is ugly, spoils the scenery (1) and may adversely affect the tourist industry (1). It is in the Peak District National Park and could conflict with ideas behind the Park (1) as it will not help to preserve the natural landscape (1).

Yes: It is close to a quarry for limestone which is the main raw material in cement – this will help to keep the costs down (2). Transporting the cement will be easy as there is a railway straight to the works which will also help to keep the lorries off the roads (2): this should be good for the environment (1). Workers could come from the nearby settlements such as Bradwell (1).
Or any other relevant point
4ES

7. (a) *1 mark for a valid point and 2 marks for a developed point.*

Yes: A population census is expensive to carry out so an ELDC would be unable to afford to do it thoroughly (1). Many people in the country are illiterate, so they could not fill in forms (2). In ELDCs, due to poverty, there is more likely to be civil unrest making it dangerous to carry out a census (1). A significant number of the population may be nomadic and therefore difficult to find (1). Fluctuating populations are difficult to record.

No: Although literacy % is low, there will be sufficient educated people in the country to conduct a census (2). Recent improvements in transport and communication have made it easier to gather census information (1). Any census can give a reasonable estimate of numbers. (1)
Any other reasoned argument for NO
3ES

Geography
General Level 2007 (continued)

7. (continued)

   (b)   • Line graph (1)
        • shows trends through time (1)
        • shows rate of change by steepness of line (1)
     • Bar graph (1)
        • good for comparing actual amounts (1)
        • can be enhanced by colour (1)
     • Pictogram (1)
        • could use bars shaped as people (1)
        • to reflect the subject of the graph (1)
        • maximum 1 mark for straight lifts
     4ES

8. Answers could include:
USA is wealthier (1) with a GNP nearly seventy times as big as that of Haiti. (1) Infant mortality is much lower in the USA (1) due to better living standards/ more hospitals/doctors (1). Education is much better in the USA (1) with almost all adults able to read and write (1) – almost double Haiti's rate (1). Life expectancy is higher in the USA (1) due to the better living standards/diet (1). USA is an EMDC whereas Haiti is an ELDC (1).
4ES

9. *1 mark per valid point, 2 marks per expanded point.*
Answers might include:
USA, Japan and Europe are the world's leading economic countries (1). The richest countries purchase the most oil (1). These countries have lots of industries which need oil (1). Many people in these EMDCs have cars which need oil to run (1). Japan does not have its own oil so will have to buy oil in (1).
4KU

10. *1 mark per valid point, 2 marks per expanded point.*
Answers might include:
Houses and businesses would have to be rebuilt (1). Electricity and water supplies would need to be properly restored (1). People may need financial help to rebuild tourist hotels and facilities: important source of income (2). Farming equipment needs to be replaced to ensure long term food supplies (2) and for the country to be self-sufficient (1).
Any other valid point.
4KU

Geography
Credit Level 2007

1. (a) *One mark for each correct answer:*
    A Docks
    B Newport on Tay
    C Railway Bridge **3ES**

   (b) *One mark for a simple point, two marks for a developed point. Max 1 for grid references.*
Answers might include:
The undulating nature of the landscape means that the A914 and A92 need cuttings and embankments (2). The roads have had to avoid the higher steeper ground (1) such as grid square 4126 (1). Most settlement is along the coast where there is low land for building (1) and where there are sandy beaches which may have attracted tourist development (1). There is also a jetty for fishing boats due to coastal location (1). Where there is gently sloping lower ground farming may be arable or mixed (1), whereas on the steeper slopes and higher ground livestock grazing will occur due to the difficulty of using machinery (2). Woodland is either grown on land which is too high and cold for crops such as 415265 (1) or steep slopes like 424254 (1). **4ES**

   (c) *Need to mention both benefits and problems for full marks. 1 for simple point, 2 for developed point.*
Benefits: New industrial estate provides jobs for local people (1). New dual carriageway improves communications and speeds up traffic flow (1). Quarry is now disused so less noise/dust pollution for locals (1). Two new schools enable growing population to be educated (1).

Problems: Increased housing means farmland is lost (1) and the quality of the environment has been spoiled (1). [The lake at 443325 has been drained removing a recreational facility (1).] The new dual carriageway means that there is more traffic and air pollution from this (1). Noise pollution/visual pollution (1). **5ES**

   (d) *Accept yes/no answers*
1 for simple point, 2 for development point.
Yes: 3930 has very little open space whereas Gauldry is in the countryside (1) with footpaths giving access to nice walks in the woods and on the hills nearby (2). 3930 has 3 A-class roads and many others so it will be noisy, whereas Gauldry will be quiet and peaceful with only a few minor roads (2). Also Gauldry will be much less polluted (1).

No: At present he has easy access to work, 1 km approx (1) whereas from Gauldry he has a much longer journey (1). He will have to drive 4km by minor road to reach the A914 and still has another 6 km to drive to work (2). Or, he will have a slow drive through the built-up area of Newport (1). Also he has to pay a toll at the bridge (1). In 3930 he will have easy access to high order services in the CBD, whereas

1. (d) continued

Gauldry only has a few low order services such as a pub and post office (2). 5ES

(e)
- Count storeys of buildings (1)
  - To test theory that building height increases towards centre of town (1)

- Pace out lengths of buildings with different functions (1) or record land use on a map (1)
  - Provides original data (1) which can be used to calculate percentages of land uses in different parts of the transect (1)

- Draw field sketches (1) or take photos along the transect (1)
  - Enables comparisons to be made between buildings in different places (1). They can be annotated to emphasise particular features (1)

- Do a traffic count, recording the amount of traffic that passes in a given time (1) and a pedestrian count (1)
  - To compare how busy it is at different points along the transect (1)

- Or any other valid techniques and justifications. 5ES

(f) *No mark for description.*
It is close to a dual carriageway which provides good access for vehicles carrying parts/raw materials and distributing finished products (1) and also provides good access for the workforce (1). It is on the edge of the city where land is cheap (1) and there is room for expansion (1). The land slopes gently so it is not difficult to build on (1). It is close to housing areas so there is a labour force nearby (1). There will be a big market in the city of Dundee (1).
4KU

2. Well annotated diagrams could obtain full marks. For full marks, answer must refer to both hanging and main valley. *Maximum of 1 mark for simple reference to plucking and/or abrasion.* 4KU

Answers might include:
A glacier moves down a main valley which it erodes (1) by plucking (1), where the ice freezes on to fragments of rock and pulls them away (1) and abrasion, where rock fragments which are embedded in the ice scour the landscape (1). The main valley is made deeper and wider (1). A smaller glacier moves down a tributary valley (1) and erodes it less deeply (1), leaving, after the ice has melted, a smaller valley whose floor is some height above the floor of the main valley (1). The answer could also refer to the shapes of the valleys before glaciation (1).
4KU

3. X (1)
Set X shows a lower temperature than set Y, which matches the position of Bristol in the cold sector of the depression (1). The high wind speed in set X fits with the closely spaced isobars at Bristol (1). The alignment of the isobars and the anticlockwise movement of air round a depression suggest that Bristol will have a SW wind as shown in set X (1). The heavy rain in set X is typical of the cold front which is very close to Bristol (1). Fronts also cause extensive cloud cover, matching the 7 oktas in set X (1). Cumulonimbus clouds, as shown in set X, are typical of a cold front where warm air is pushed up rapidly (1). 5KU

4. (a) No marks for choice of physical and human causes. 1 mark for a simple point, 2 marks for a developed point. To gain full marks, answer must refer to both physical and human causes.
- If 'unreliable rainfall' is chosen from physical:
  Some years are wetter and some are drier than others (1). If areas have several dry years in succession, crops will not grow (1). There are no roots to hold/bind the soil (1) so it can be blown away by the wind (1) or eroded by people/animals trampling over it (1).
  Any other valid point.

- eg if 'population increase' is chosen from human:
  If population increases through a rise in births and immigration, more land will be needed for crops (2). There will be less fallow land because more food is required to feed the people (1). Overcultivation occurs and the soil loses fertility (1). Fewer crops can grow so there is no vegetation cover to bind the soil together so erosion takes place (1) and the desert spreads (1).
  Any other valid point. 4KU

(b) *1 mark for a simple point. 2 marks for a developed point.*
- Irrigation allows crops to be grown (1). Crop roots then bind the soil together, preventing it from being washed/blown away (1). Also, wetter soils are heavier and harder to remove (1).
- Plant more trees because the roots hold the soil together (1). Trees can improve soil fertility (1), can provide shelter for crops (1) and slow down and/or divert winds which can damage crops (1).
- Terraces in hillsides can help to trap water (1), preventing soil being washed away (1) so more crops can be grown which then holds the soil together (1).
- Stone lines built along contours on sloping farmland (1) trap water after rain so the soil is not easily washed away (1). Soils remain deep behind stones and crops will grow (1).
- Any other valid point. 4KU

Geography
Credit Level 2007 (continued)

5. *1 mark for a valid point. 2 marks for a developed point. Answer must explain changes.*
Answers might include:
Shopping centres are under cover so they provide more pleasant shopping environments (1). It is also easier for shoppers if all shops are in a compact area rather than spread out along a street (1). Shops are closing down due to more out-of-town shopping centres (1) which are more accessible by car and convenient to get to (1). The buildings left are used for other purposes – usually entertainment (1). High parking charges discourage motorists from taking cars into the CBD (1), so reducing traffic congestion (1). One-way streets to allow traffic to flow more freely (1). Pedestrianised areas increase safety for shoppers and reduce air pollution (2). New buildings are often multi-storey because there is a high demand for land in CBD and no room to expand sideways (2). 5KU

6. *1 mark for a valid point. 2 marks for a developed point.*
Human: Human factors create more problems than physical factors in this sort of area because farmers are so dependent on EU subsidies (1) and so any changes made to the level of subsidy would have a big impact (1). Inverlochlarig and other sheep farms will have been hit by falling livestock prices in recent years (1) as well as rises in fuel prices (1). The farm's activities may be restricted by new conservation regulations for upland areas (1) and there will be difficulties caused by walkers leaving gates open and disturbing sheep particularly at lambing time (2). The relatively long distances involved in sending goods to market will restrict the farm's profits (1). Or any other point.

Physical: Physical factors will have a bigger impact because the steep slopes will prevent the use of machines and so restrict the potential for arable farming (1) and the high levels of rainfall in the mountains will make the ground boggy and difficult to farm (1). In winter the higher slopes will be cold and unsuitable even for sheep and snowfall may well cut the farm off for several days at a time (1). Soils are likely to be thin and infertile making farming tricky (1). It will take the farmer a long time to round up sheep for dipping, shearing etc because of the large size of a hill sheep farm and the mountainous nature of the landscape (2). Or any other point. 6ES

7. *1 mark for a simple point. 2 marks for a developed point. Answers must be explanatory.*

Advantages: There are 650 jobs which will help the local economy (1) because the workers will be able to spend money in surrounding businesses (1). Redundant workers from the old china clay industry may now be employed there (1). The work is mostly permanent, not seasonal like many tourism-related jobs (1). The old quarry may have been a blot on the landscape and the Eden Project will have smartened it up (1). Many visitors will need accommodation and this will boost the income of local guest houses and hotels (1). There are also educational opportunities for local children (1).
Any other valid point.

Disadvantages: There are 1·8 million visitors each year and most of them will come by car so this will increase congestion on the roads, while exhaust fumes will cause more pollution (2). The Eden Project is not on a motorway or dual carriageway, increasing the pressure of traffic on smaller A-class roads (1). Parking spaces for 5000 vehicles will take up a lot of room and may spoil the look of the area (1). With such a large number of visitors, it may encourage other new developments such as hotels and restaurants to mushroom around it which would have a negative effect on the environment (1).
Any other valid point. 6ES

8. *Max 2 for simple descriptive links between population density and physical feature, e.g. 'It is sparsely populated in mountains and areas of low rainfall and densely populated where there is coal' (2).*
The areas with low rainfall are sparsely populated because water is needed to irrigate crops for domestic use (1). Mountain areas are sparsely populated because they are too cold and the soils are too thin for crops to be grown (2). Northern Canada is sparsely populated because at these latitudes there are long dark winters and it is too cold to grow crops (2). The East is densely populated because there is sufficient rainfall for agriculture and for domestic water supplies (1). Also there is coal here providing employment in mining (1). Other industries will use the coal for power and attract people into the area for the jobs they provide (2). In the west there are some densely populated areas close to HEP stations which provide power for industry (1). 6ES

9. (a) *1 mark for a valid point. 2 marks for an expanded point. Max 1 mark for straight lifts. Accept agree/disagree answers.*

Agree: The tourism industry brings much-needed money into ELDCs (1), which should improve the standard of living of poor people (1). It will provide jobs (1) which don't require a lot of skills (1) for locals. Money can help to support projects such as nature reserves (1).

Disagree: Locals can't use the beaches and facilities (1). Employment is only seasonal, during the European winter (1). Prices are low due to competition, which limits the amount of money gained (1). All tourist spending is in the coastal areas so people inland don't benefit (1).
4ES

(b) *Techniques must be linked to the relevant data.* Answers might include:

- Bar chart for labour force and tourism employment (1)
  - Comparison can be made between them (1)
- Flow map for European visitors (1)
  - Thickest line will show where most came from (1)
- Pie chart for GDP figures (1)
  OR
  Divided bar graph for European visitors (1)
  - Data in percentages and these techniques show proportions (1)
- Or any other valid techniques/reasons.
4ES

10. *1 mark for a simple point and 2 marks for a developed point.*

For example: Primary Employment %
In ELDC's many people are engaged in agriculture needing to grow their own food to survive whereas in EMDCs people buy their food with the money they earn working in secondary and tertiary jobs (2). In EMDCs wealth is based on the development of industry so more people will be in secondary and tertiary employment (1). In EMDCs extracting raw materials is done mainly by machinery, whereas in ELDCs much of the work is done by hand requiring more workers (2).
6KU

Geography
General Level 2008

1. (a) *All 4 correct – 3 marks 2 or 3 correct 2 marks 1 correct – 1 mark*

Grid Reference/ Name of feature	Letter	
Angel's Peak	D	Pyramidal peak
Lairig Ghru	C	U-shaped valley
Loch Coire an Lochain	B	Corrie
Loch Einich	A	Ribbon lake

(b) *Credit should be given for relevant diagrams Fully annotated diagrams could gain 3 marks.* Possible answers might include: Snow collects in hollow then turns to ice (1) this ice eroded the mountain on all sides (1) creating corries (1) these got progressively bigger and their back walls were eroded back towards each other (1) frost shattering produced a sharp peak (1)
3 KU
Maximum 1 mark for a list of processes eg abrasion, plucking, freezing, thaw (1)

(c) *For full marks both points of view should be mentioned. No marks for grid references. Mark 3:1, 2:2, or 1:3* Possible answers might include: Settlement Encouraged: flat land along the valley floor making building easy (1) route way through valley (1) beside river for water supply (1) bridging point on river (1). Limited: surrounded by steep slopes making building difficult (1) possible flooding from the river (1) expansion restricted by the steep slopes (1). Accept reference to forestry (1).
4 ES

(d) *Accept yes/no answers. No marks for choice. Max 1 for list of activities/features.*

Possible answers might include:
Good: surrounded by steep slopes for rock climbing, abseiling (1), loch available for water sports eg sailing, canoeing (1), forest trails for orienteering (1), access to ski slopes (1), good training area for mountain rescue teams (1).
Bad: very isolated (1), few roads into the area (1), inaccessible at certain times of the year (1), lack of accommodation (1).
4 ES

(e) *Maximum of 1 mark for a list (1).* Possible answers might include: Beauty of the countryside destroyed by main road through the area (1), scenery affected by the ski tows (1), beauty taken away by caravan and camp sites (1), litter dropped by tourists (1), views obscured by forestry (1), access

Geography
General Level 2008 (continued)

1. (e) continued
restricted by farms, forestry (1), access limited
by narrow, country roads (1), local wildlife
disturbed by tourists walking on the hills (1),
noise from clay pigeon shooting frightens
animals (1), deer killed on roads (1), destruction
of local animal habitat by ski slopes (1), rare
plants and small animals affected by footpath
erosion (1).
Accept conflicts between visitors and farming.
4 ES
Mark 1:3, 2:2, 3:1

(f) Techniques might include:
taking photographs (1), drawing field sketches
(1), questionnaire to walkers and cyclists who
use the path (1), visit the area and observe (1),
interview locals who stay along the route (1),
interview park ranger (1). Interview
shopkeepers (1).
*Maximum of 1 mark for mention of same
technique.*
Reasons could include:
photographs can be compared if taken at
different times to show any changes (1), walkers
can give you up to date information (1), and if
they use the path frequently can indicate the
damage at different times (1), park ranger would
have first hand knowledge of impact (1).
Shopkeepers can comment on economic impact
(1).
4 ES

2. *One mark per valid point, two for developed
statements. Credit suitably labelled diagrams.*

Delta: this is formed when the river
deposits sediment (1) because the current slows
down when it meets the sea and no longer has the
power to carry all its load (2) the sediments build
up to form islands (1) and the river flows between
them in a braided channel (1).
Flood Plain: flood plains are formed when a river
bursts its banks and it deposits silt/alluvium on the
surrounding land (1). The edges of the flood plain
can be made wider due to erosion on the outside
bends of the meander (1) and as these meanders
move with time, different parts of the flood plain
are widened (1).
Credit references to levees and channel deposits.
Ox-Bow Lake: the current of the river erodes the
outside bends of a meander (1), this causes the neck
of the meander to get narrower (1) until eventually
the river breaks through and leaves the old meander
abandoned (1) the ends of the meander are sealed
off due to deposition (1) leaving a crescent shape or
ox-bow lake (1).
3 KU
There is no credit for reference to meander without
appropriate detail.
Credit 'deposition on neck of meander' (1)

3. *Max 2 marks for correct description of weather.*
No credit for reference to cloud cover or wind
direction.

Answers might include:
Snow was forecast (1) which would have made the
pitch impossible to play on (1) and would have
blocked the roads leading to Pittodrie (1).
Temperatures were to be below freezing (1) so the
pitch would have been frozen (1) and consequently
dangerous for the players (1). Strong winds would
have made it difficult to play football (1) and would
have caused the snow to drift, blocking access roads
(1).
4 KU

4. (a) 2 marks for placing the two lines correctly.
1 mark for labelling.
3 KU

(b) Answers might include:
A: enforce laws banning dumping: this
would reduce the amount of leaks from oil
tankers (1); force companies to find safer ways
of getting rid of waste (1); perhaps help to
encourage more recycling (1).
B: ensure sewage is treated: would benefit
marine life (1) and make beaches safer for
people (1), easier to monitor (1).
3 ES
Accept negative points about other measure. (eg
enforcing laws would be difficult (1))

5. *Mark 1:3, 2:2, or 3:1*
No marks for naming climatic regions.
Max 2 marks for description of each climate.

Answers might include:
X: extremely cold temperatures (1) mean that most
of the year the ground is frozen and cannot be
cultivated (1), growing season is too short for
farming (1), working outside is difficult in such cold
temperature (1) with danger of frostbite (1).
Y: climate is extremely dry (1) making agriculture
difficult (1) water is a basic requirement of life so
people cannot live here easily (1) very high
temperatures can cause heatstroke (1).
4 ES

6. *No marks for choice.*
Maximum 1 mark for description or straight lift.

Dairy Farming: close to market for easy transport
(1), low cost of transport on a daily basis (1), flood
plain good for grazing (1), grass rich on flood plain
(1), flat land suitable for dairy cattle as they do not
like slopes (1), mild temperatures suitable for cows
(1), enough rainfall for good grass (1).
Accept negative points.
Arable Farming: adequate rainfall for growing
crops (1), good flat land for use of machinery (1),
fertile soil saves on using fertilisers (1) close to
market for produce (1).
4 ES

7. *1 mark per valid point, 2 marks for expanded point.*
At least two features must be explained.

Answers might include:
"Close to main roads and motorways" means quicker/easier transport of goods (1) which cuts costs for companies (1), it also makes it easier for workers to travel (1).

"Near edge of city" means land is cheaper away from CBD/inner city (1) but site is still close enough to town for a labour supply (1) it has a pleasant environment away from noise/pollution/dereliction in old industrial areas (1) "Landscaped with grass, trees and shrubs" creates a pleasant environment which helps in the recruitment of skilled workers (1) and also promotes a positive image of the company (1). "No chimneys" – modern factories tend not to use coal/steam power (1) but use electricity (1). "Spacious site on flat land" – means room for expansion (1) and flat land is easy to build on (1). "Large areas of tarmac surface" – parking areas for workers who nowadays mostly travel by car (1) parking areas/turning areas/delivery bays since most transport for modern industry is by lorry (1).
4 KU

8. *No marks for choice.*
Accept negative points about the alternative.

Route A
Good farmland would not be lost (1), only a few trees would need to be cut down (1), does not have to pass through any built up areas (1) so residents would not be worried by noise or accidents (1), there are no other roads nearby so no tunnels or bridges have to be built (1).

Route B
This would be shorter so not as expensive to build (1), would not need to build cuttings as land is flat (hills along Route A) (1), or cut down trees which destroy wildlife habits (1), would provide good transport links for farmer taking products to market (1), could be used by industrial estates to import raw materials and transfer finished products (1) reducing transport costs (1).
ES 4

9. (a) *Two factors must be mentioned for 4 marks.*
Mark 2:2, 3:1 or 1:3.
Accept negative points.

Answers might include:
Diet: improved diet leads to better health (1) so people live longer (1), balanced diet leads to fewer dietary diseases (1) eg Rickets (1), a balanced diet also reduces the risk of heart disease (1) so increasing life expectancy (1).

Medical Advances: new technology can provide an earlier diagnosis (1), earlier identification allows faster treatment of illnesses (1) reducing the death rate (1), new drugs cure or prevent illness so improve life expectancy (1), advances in treatment of heart disease eg heart bypass surgery allow people to survive longer (1).
4 KU

9. (b) *1 mark for relevant technique: Line Graph (1), Bar Graph (1)*
Any other valid technique with reason.

Reasons could include:
line graph shows trend over time (1) can show rate of change by steepness of the line (1) bar graph shows good visual comparison (1) can be enhanced by colour (1).
ES 4

10. (a) *1 mark per valid point, 2 marks for developed point.*
Mark 3:1, 2:2 or 1:3. Advantages and disadvantages must be mentioned.
Max (1) for straight lift.
Possible answers might include:

Advantages: money can be used to buy seeds or new machinery (1) so that food supply improves (1), improved trade links with EMDCs (1) would allow ELDC to fund projects like HEP schemes (1).
Disadvantages: money may have to be paid back with interest (1) which may result in more debt for ELDC (1), ELDC may not have choice about where to spend money (1), ELDC may have to buy replacement parts for machinery from donor country (1) which means they cannot shop around for the cheapest prices (1).
4 ES

(b) *1 mark per valid reason, 2 marks per expanded point.*
Accept yes/no answers.

Possible answers may include:
Yes: better education about family planning would lead to better food supply (1), educating children may help them get better jobs when they grow up (1), being able to read and write means that people can understand leaflets about family planning/health care (1), earning more money when the children grow up would mean easier living conditions (1) and could help support parents (1).

No: takes some family away from working on the farm (1), needed to grow essential crops/food (1), means traditional way of life will be lost (1), people may be happy with present way of life (1).
3 ES

11. 4 *marks for six correct, 3 marks for four or five correct, 2 marks for two or three correct, 1 mark for one correct.*

UK
• Exports mainly manufactured goods. B
• Agriculture employs 2% of population. E
• High GNP per capita. F

Geography
General Level 2008 (continued)

11. continued

India

- Quotas on exports to Germany and France low.
  A
- Agriculture employs 66% of population.
  C
- Energy consumption per capita is low.
  D

4 KU

Credit Level
Geography 2008

1. (a) *1 mark per valid point, 2 marks per expanded point.*
Both river and valley must be referred to for full marks.
Max (1) for grid reference.

Possible answers might include:
The River Aire flows in a general NW to SE direction (1) and has a winding course (1) with several meanders (1) eg at 085417 (1); the feature at 063427 may be a cut off (ox-bow lake) (1); the river is joined by several tributaries eg at 043448 (1). Gradient is gentle (1).
The valley floor is wide (1) up to 1km wide in grid square 0841 (1) and flat (1). The sides of the valley are fairly steep (1) and rise from about 90 metres on the valley floor to over 300 metres eg in 0743 (1).
4 KU

(b) *A well annotated diagram could earn full marks.*

This is a V-shaped valley eroded by the river in its upper course (1) created by down cutting/vertical erosion (1) by corrosion and hydraulic action (1); the exposed sides are weathered (1) eg by freeze-thaw action (1); particles are moved down the slope by the movement of rainwater/gravity (1) and transported away by the stream (1) eg by saltation and traction suspension (1).
4 KU

(c) *1 mark per valid point, 2 per expanded point.*
Answers referring to only advantages or disadvantages can achieve full marks.

Advantages: land to the north of the farm is fairly gently sloping and could be suitable for pasture (1) and even for cultivation using machinery (1). There is access to services in nearby Ilkley (1) and to markets via the A65 primary route (1). Land is well drained because of slope (1).
Disadvantages: there are several footpaths in the area, and the Dales Way nearby, visitors/walkers could be a nuisance (1) location on the outskirts of town could cause vandalism/dumping problems (1) some of the land is very steep, especially to the south of farm buildings making use of machinery difficult (1). Woodland may be habitat for pests eg rabbits (1) land is high, >260 metres so temperatures may be low and it would be hard to grow crops (1).
4 ES

1. (d) *No marks for grid references.*
1 mark per valid point, 2 marks for developed point.

Possible land uses likely to be mentioned would include:

Farming, urban expansion eg south east of Ilkley, local recreation (golf course) and tourist recreation (walking on Dales Way), hotel, parking, sites of historical/cultural interest (eg Pancake Stone).

Possible answers might include:
Conflict between farmers and tourists because farmers complain that tourists damage walls and fences (1), leave gates open allowing animals to stray into danger (1), allow pet dogs to worry sheep (1) and tourists object to farmers' attempts to restrict access (1).
Conflict between urban areas/developers and farmers over spread of urban areas/new housing onto farmland (1) hampering farmers and even putting them out of business (1).
Conflict between either of these groups and recreational users such as golfers (1).
Conflict between golf course and walkers over access (1).
Conflict between conservationists and walkers over potential damage to historic sites such as Pancake Stone (1).
Conflict over traffic congestion between local residents/road users and tourist traffic at busy sites on narrow roads eg near the car park and viewpoint in grid square 1346 (2).
Accept any other relevant points
6 ES

(e) *1 mark per valid point. 2 marks for a developed point. Marks only for differences*

Answers may include:
0440 is an area of newer housing, 0641 an older town centre/inner urban area (1).
0440 mainly suburban residential, 0641 greater variety – town centre, shops, offices and industry as well as housing (2).
0440 varied street pattern including crescents and cul-de-sacs, 0641 less varied (linear/rectangular/grid-iron) (1).
0440 has smaller buildings (houses), 0641 has large buildings including factories/industry (1).
0440 limited amount of traffic, less noise pollution, 0641 many main roads, railway (and station), bus station – more noise and pollution (2).
0440 more open spaces/access to countryside including nature reserve, 0641 less open space and poorer quality environment (2).
Accept other relevant points.
5 ES

1. (f) *1 mark per valid point, 2 for an expanded statement.*
Good access to network of main roads for cheaper/easier transport of goods (1). This is an important location factor in cutting costs (1); water supply from river (1); flat land for building (1); beside town for labour supply (1). Area around town has attractive scenery/recreational opportunities which help to attract workers (1).
Accept other valid points.
5 ES

2. *One mark per valid point, two for a developed statement.*
For full marks two features must be mentioned. Credit suitably labelled diagrams.

Possible answers include:
This landscape was formed by glacial deposition (1), the terminal moraine was formed when a glacier picked up moraine and transported it (1), when the glacier melted the moraine was deposited at the snout of the glacier, marking the furthest point it reached (2).
Beyond the terminal moraine, the outwash plain was formed by melt water streams (1), this transported fine material such as sand and gravel from the ice (1), before sorting and depositing it (1).
Drumlins are small hills made of boulder clay which are moulded by the ice as it passes over (1) the boulder clay is formed beneath the ice as rocks and other debris are ground up and smashed by the erosive effect of the ice sheet/glacier (1).
5 KU

3. *Accept yes/no answers.*
Maximum 1 mark for simple descriptions/list of weather characteristics
No credit for reference to a cold front. If reference to warm front, movement must be mentioned.

Yes: High pressure/anticyclone is covering Britain (1). This could bring settled weather for the whole week (1). Skies will be clear and sunny due to the lack of any fronts (1).
Isobars are well spaced so winds will be light (1). Wind will blow from south bringing warm weather (1), since it is summer temperatures will be hot under the clear skies (1).
No: The centre of the anti-cyclone lies to the east of Britain, suggesting it might move away to the east (1). A warm front is approaching from the west (1). This will bring cloudy conditions and heavy rain due to condensation of moisture in the rising air at the front (2). The isobars will become closer together causing high winds (1).
The high pressure could cause a heat wave (1) with dangers of dehydration for children (1) and sunburn in the sunshine (1). There is a possibility of thunderstorms after a spell of hot weather (1).
5 ES

Geography
Credit Level 2008 (continued)

4. *No marks for straight lifts.*
Accept yes/no answers.

Yes: Carbon dioxide levels in the atmosphere will increase (1) causing global warming (1), raising sea levels by melting of ice caps (1) with world-wide flooding of lowland areas (1).
The valuable hardwood timber is mostly exported, so the world benefits from these resources (1).
Since it has the largest number of plant/animal species of any natural region its destruction will have a major impact on world biodiversity (1).
Since many medicines are derived from plants, losing them will seriously affect research into cures for diseases throughout the world (1).
Cattle ranching and plantation agriculture provide cheap food for export to other countries (1)
Many of the countries exploiting the forest are multinationals, so profits leave the country (1) and Brazil does not benefit much economically (1).
Exploiting the mineral resources, will if they are exported, provide raw materials for many countries, not just Brazil (1).
No: The exploitation of resources like timber and minerals provides jobs for Brazilians and improves the economy of the country (2).
It is only Brazil's wildlife that is affected, not the rest of the world (1).
The HEP provides energy for Brazil's industry (1).
The fact that Indians are forced to live on Reserves only affects people in Brazil (1).
The building of the road means that Brazilian settlers can gain access to virgin forest, providing them with a livelihood (1), and also gives access to resources for Brazilian companies to exploit (1).
6 ES

5. *(a) One mark for a valid point. Two marks for a developed point.*
1 mark for description. For full marks both zones must be referred to.

Possible answers might include:
Zone 1 19th C:
In the inner city tenements/terraced housing were built to save space (1), because this zone is close to the CBD where land is expensive (1) this allowed high housing (population) densities (1), houses were close to industry because people had to walk to work (1), little open space or gardens as land was scarce (1).

Zone 3 late 20th C
Because it is on edge of town where land is cheaper there will be low housing density (1), houses are larger, detached or semi detached with back and front gardens and garages (1), newer housing so better planned layout with cul-de-sacs and crescents (1).
6 KU

5. *(b) At least two techniques must be described.*
Maximum of three marks if no reasons are given or if reference is made to only one technique. Do not credit the same reason twice.
Mark 2:3 or 3:2.

Possible answers might include:
Comparison of old and new photographs (1) these could be displayed side by side to highlight changes in land use (1) and differences in the amount of open space, building heights and street layouts (1).
Photographs could be annotated to show changes (1).
Looking at old and present day maps (1), saves the need for time consuming fieldwork and would show changes in land use (1) and differences in the amount of open space (1), services available then and now (1).
Fieldwork in CBD could record building age, height and function (1), able to record present day land use (1), would be able to compare this with old records, photographs of area (1).
In any of the above give credit for old materials obtained from library/planning offices.
5 ES

6. *(a) For full marks both benefits and problems must be mentioned.*

Benefits: Economically such developments create more jobs, so cut unemployment (1) meaning people have a higher standard of living (1), the multiplier effect is likely to kick in (1) so tradesmen and other local firms will get more business from new factories (1), while shops and other services will benefit from local people having more money to spend (1). The economy of the local area will also benefit from increased tax revenue (1), with more money available for spending by the local authority on public services and the environment (1). Socially, more jobs means people are less likely to leave the area (1), avoiding problems such as depopulation and an ageing population (1). Higher employment levels and higher standards of living are usually associated with a decrease in social problems (1), including crime, vandalism and drink and drug abuse (1) and can reduce stress in family life (1).

Problems: An increase in industrial traffic and movement of workers causing traffic congestion and noise and air pollution (2), firms may want to occupy Greenfield sites with a resulting impact on the environment (1) and if the demand for houses increases, prices may rise rapidly which will cause disadvantage to the less well paid and to first time buyers (2). 6 KU

(b) Mark 3:2 or 2:3
The information in table 1 could be shown by a bar graph (1) or a pie chart (1).
Answer must link technique to appropriate table except for bar graph.

Bar Graph: A bar graph is good at showing

individual totals for each category (1). Bars are side-by-side for easy comparison (1) and can be coloured to enhance the presentation (1).

Pie Chart: only appropriate for table 1
Candidates must refer to figures being converted to percentages before crediting other statements.
Pie chart is good at showing proportions (1), colour can be used to highlight the different segments (1).
The information on table 2 could be shown by a line graph (1) or a bar graph (1).

Line Graph: Line graphs are good at showing change over time (1) showing not only rising and falling totals clearly but also the rate of change (1) allowing easy comparison between different periods (1).
If bar graph used twice as technique, only one mark.
5 ES

7. (a) *For full marks at least two stages must be described.*

Possible answers include:
Birth Rates: In stage 1 these are high (over 40/1000) (1) remaining high in stage 2 (1) they fall sharply in stage 3 to around 10 per 1000 (1) levelling out and remaining low in stage 4 with one or two small baby booms (1).

Death Rates: These were high and fluctuating in stage 1, between 43/1000 to 46/1000 (1) falling sharply in stage 2 to around 11/1000 (1) remaining low in stages 3 and 4 (1).

Total Population: This remained low in stage 1 and into stage 2 (1) halfway through stage 2 total population started to increase (1) total population increased dramatically in stage 3 (1) levelling out in stage 4 (1).
4 ES

(b) Possible answers include:
Nigeria: In Nigeria there is little contraception available so birth rates are high (1), high numbers of births to ensure some children survive (1), children are needed for an income (1) and to look after parents in their old age (1) as no government pensions (1).
Death rates are falling due to Primary Health Care (1) foreign aid (1) availability of medicines from developed countries (1) high birth rates and falling death rates means a large natural increase (1).
Any other valid point.

UK: Low birth rates and low death rates mean very low natural increase (1). Low birth rates results from better medical care of mothers and babies (1), so most babies survive (1) so lower birth rate (1), low birth rate from use of contraception (1), birth rates low from later marriages (1), children are expensive (1), women want careers rather than children (1).

7. (b) continued

Low death rates due to medical advances (1), good living conditions (1), National Health Service (1).
Accept other valid points.
4 KU

8. (a) *1 mark for straight description of graphs.*
1 mark per valid difference, 2 marks per expanded point.

Possible answers might include:
Exports mainly manufactured goods – 97% but imports quite a lot of raw materials (1) like food and oil (1), doesn't seem to export any raw materials (1). 38% of imports are raw materials (1) only possible 3% of exports are raw materials (1).
3 ES

(b) *1 mark per valid point, 2 marks per expanded point.*

Possible answers might include:
Japan has very little in the way of natural resources of her own (1) so has to import them (1) eg oil and timber (1), relies on making goods using bought resources (1) to be able to make money (1), will not pay highly for raw materials but can sell electrical goods for high prices (1) so that businesses make large profits (1).
3 KU

Official SQA answers to 978-1-84372-632-6
2004–2008

Official SQA answers to 978-1-84372-632-6
2004–2008